LIKE HEALS LIKE . . .

The ancients believed that a liver ailment in a human could be helped by eating animal liver, and modern medical research is exploring the very real basis for this theory, showing a wide range of benefits of organ and glandular extracts. Dr. Murray explains the therapeutic uses of thyroid, liver, adrenal, cartilage, aorta and other extracts and how you can determine if they have a place in your health program.

ABOUT THE AUTHOR

Michael T. Murray, N.D., one of the world's leading authorities on natural medicine, is a graduate and faculty member of the Bastyr College in Seattle, Washington. In addition to maintaining a private medical practice, Dr. Murray is an accomplished writer, educator and lecturer. He is coauthor of *A Textbook of Natural Medicine,* the definitive textbook on naturopathic medicine for physicians, and *The Encyclopedia of Natural Medicine,* a reference book of naturopathic medicine for consumers. He has also written several other books, including *The Healing Power of Foods* and *Natural Alternatives to Over-the-Counter and Prescription Drugs.*

GLANDULAR EXTRACTS

WHAT YOU MUST KNOW

HOW NATURAL CONCENTRATES STIMULATE ENDOCRINES, PROVIDE ENZYMES AND PROMOTE CELLULAR REPAIR

by Michael T. Murray, N.D.

Keats Publishing, Inc. New Canaan, Connecticut

Glandular Extracts: What You Must Know is not intended as medical advice. Its intent is solely informational and educational. Please consult a health professional should the need for one be indicated.

GLANDULAR EXTRACTS: WHAT YOU MUST KNOW

ISBN: 0-87983-611-3

Printed in the United States of America

Good Health Guides are published by
Keats Publishing, Inc.
27 Pine Street (Box 876)
New Canaan, Connecticut 06840-0876

Contents

Introduction 7
Pancreatic Extracts 11
Thymus Extracts 15
Spleen Extracts 19
Liver Extracts 22
Adrenal Extracts 26
Thyroid Extracts 28
Cartilage Extracts, Chondroitin Sulfate and Glucosamine 34
Aorta Extracts 38
Final Comments 42
References 43

INTRODUCTION

For almost as long as historic records have been kept, glandular therapy has been an important form of medicine. The basic concept underlying the medicinal use of glandular substances from animals is that "like heals like." For example, if your liver needs support or you are suffering from liver disease, then you may benefit from eating beef liver. Modern glandular therapy, however, primarily involves the use of concentrated glandular extracts.

What is a gland?

A gland is a secretory organ. The internal secretory organs of the body are called endocrine glands. These ductless glands secrete hormones directly into the bloodstream. The glands which are known to have endocrine function include the pineal, pituitary, thyroid, parathyroid, thymus, adrenal, pancreas, and gonads (testes or ovaries). Although they are not technically glands, it is common to refer to other organs of the body as glandulars when they are used in glandular therapy. For example, tissue extracts of heart, spleen, prostate, uterus, brain, and other tissues are often used in glandular or organotherapy.

Are glandulars effective?

Science has confirmed that certain glandular preparations and hormones are quite effective when taken orally. It is well established that a number of glandular preparations are effective orally because of active hormone or enzyme content, e.g., thyroid, adrenal cortex, and pancreatin preparations. There is also a good deal of literature support for pharmaceutical grade liver, aorta and thymus extracts, and some support for pituitary, spleen, orchic (testes) and ovarian extracts as well. However, despite this scientific backing, many people still question the effectiveness of glandular products on human health.

The true question should not be, are glandulars effective? but rather, are the glandulars currently available in the health food stores in the United States effective? Manufacturers of glandular products will ballyhoo that their method of production of a glandular extract is the most nearly ideal. However, the majority of their contentions are based on theoretical or philosophical grounds, not on firm scientific evidence or clinical results. There are no quality control procedures or standards enforced in the glandular industry. It is left up to the individual company to adopt quality control and good manufacturing procedures.

Nonetheless, I believe that many glandular preparations available in the United States marketplace are extremely effective. There are several companies which manufacture high quality glandular products.

Understanding Glandular Preparations

It is critical that consumers purchase glandular products from well-known reputable companies, as improperly processed glandular material will not produce results. The biologically active material, such as enzymes, soluble proteins, natural lipid factors, vitamins, minerals and hormone precursors are destroyed or eliminated if the product is not prepared properly.

Most glandular products are derived from beef (bovine) sources; the exception are pancreatic extracts which are most often derived from pork (porcine). The four most widely-known methods of processing are the azeotrophic method, salt precipitation, freeze-drying, and predigesting.

The azeotrophic method begins by quick-freezing the material at well below 0 degrees F., and then washing the material with a powerful solvent, ethylene dichloride, to remove the fatty tissue. The solvent is then distilled off and the material is dried and ground into a powder so that it can be placed in tablets or capsules. Although the azeotrophic method eliminates the problem of fat-stored toxins like pesticides and unhealthy metals, unfortunately it also removes fat-soluble hormones, enzymes, essential fatty acids, and other potentially beneficial materials, and traces of the solvent still remain.

The salt precipitation method involves the maceration of fresh glandular material in a salt and water. Because the salt increases the density of the water-soluble material, when the mixture is centrifuged the lighter fat-soluble material can be separated out. The material is then dried and powdered. The benefit with the salt precipitation method is that no toxic solvents are used to separate the fatty material. The down side is that most people could do without the salt in the product.

The freeze-drying process involves quickly freezing the glandular material at temperatures 40 to 60 degrees below zero degrees F. and then placing the material in a vacuum chamber which removes the water by direct vaporization from its frozen state—hence the term freeze-drying. The benefit of freeze-drying is that it produces a higher concentration of unaltered protein and enzymes as well as all of the fat-soluble components. Since the fat is not removed, it is critical that the glands be derived from livestock that grazed on open ranges, unsprayed with pesticides or herbicides. The animals must also be free from antibiotics, synthetic hormones and infection.

The final process to discuss is the predigestion method. This method employs the aid of plant and animal enzymes to partially digest or hydrolyze the glandular material. The partially digested material is then passed through a series of filtrations to separate out fat-soluble and large molecules. The purified material is then freeze-dried. This method of extraction is ideal for glandulars (such as the liver and thymus) where the polypeptide (small proteins) and other water-soluble fractions are desired.

In general, I believe that predigested soluble concentrates are the most effective. Like any other foods, glandulars must be digested to extract their nutrients. Because predigested glandulars need little or no further digestion, the body can assimilate their natural factors quickly and easily. Although my belief is based partly on theory and philosophy, there is clinical evidence which supports it.

Evidence for Intact Protein Absorption

Contrary to long-held theories that the healthy intestinal lining (mucosa) forms an essentially impermeable barrier to proteins and large polypeptides, there is now irrefutable evidence that large macromolecules can and do pass intact from the human gut into the bloodstream under normal conditions. In some instances, the body appears to recognize which molecules it needs to absorb intact and which molecules it needs to break down into smaller units. This phenomenon may help to explain the effectiveness of glandular therapy.

Numerous whole proteins, including plant and animal enzymes, have been shown in human and animal studies to be absorbed intact into the bloodstream following oral administration.[1-7] These include human albumin and lactalbumin, bovine albumin, ovalbumin, lactoglobulin, ferritin (M.W. 500,000), chymotrypsinogen, elastase and other large molecules.

Furthermore, proteins and polypeptides (small proteins) as well as various hormones that are absorbed intact from the gut have been shown to exert effects in target tissues. For example, in addition to thyroxine or thyroid hormone and cortisone, several peptide hormones are known to be biologically active when administered orally, including luteinizing hormone releasing factor and thyrotropin releasing hormone.[8,9] Even insulin has been shown to be absorbed orally under certain circumstances (e.g., in the presence of protease inhibitors or hypertonic solutions in the intestines).[10,11]

What all this data indicates is that there may be larger molecules in glandular products absorbed intact that exert beneficial effects, particularly polypeptides, which exert hormone or hormonelike action. Detractors of glandular therapy often claim that when a glandular is consumed it is immediately broken down by the body into smaller components like amino acids. However, these detractors are not up to date with current understanding of normal physiology.

Summary

From the scientific data that currently exist, there is enough evidence to support the use of orally administered glandular extracts. Consumers should choose glandular products made by reputable companies that employ established methods of manufacture.

PANCREATIC EXTRACTS

The pancreas produces enzymatic secretions required for the digestion and absorption of food. Each day the pancreas secretes about 1.5 quarts of pancreatic juice in the small intestine. Enzymes secreted include lipases, which digest fat, proteases, which digest proteins, and amylases, which digest starch molecules.

There is strong evidence that the body seeks to conserve its digestive enzymes by reabsorbing them. Several human studies have shown that when supplemental pancreatic enzymes (trypsin and chymotrypsin) are given orally, they are absorbed intact into the bloodstream.[1-5] Even more dramatic is the finding that pancreatic enzymes are not only absorbed intact from the gut, but also transported through the bloodstream, taken up intact by pancreatic secretory cells, and resecreted into the intestines by the pancreas. The existence of this circulation of pancreatic enzymes is quite similar to the recycling of bile salts by the liver.[6]

Of further interest is the possibility that oral supplementation with pancreatic enzymes may have a sparing effect on the body's own digestive enzymes, perhaps reducing the stress on the pancreas and allowing the body to redirect energy to other areas.

Pancreatic Enzymes

Lipases: The pancreatic lipases, along with bile, promote digestion of fats. Deficiency of lipase results in malabsorption of fats and fat-soluble vitamins.

Amylases: Amylases are enzymes which break down starch molecules into smaller sugars. Amylase is secreted by the salivary glands as well as the pancreas.

Proteases: The proteases secreted by the pancreas (trypsin, chymotrypsin, and carboxypeptidase) function in digestion by breaking down protein molecules into single amino acids.

The Clinical Use of Pancreatic Enzymes

Pancreatic enzyme products are popular nutritional supplements. Most commercial preparations are made from fresh hog pancreas. Pancreatic enzymes are most often employed in the treatment of pancreatic insufficiency. This condition is characterized by impaired digestion, malabsorption, nutrient deficiencies and abdominal discomfort. Pancreatic enzymes are also used by physicians in the treatment of cystic fibrosis, inflammatory and autoimmune diseases like rheumatoid arthritis and scleroderma, athletic injuries, tendinitis, etc.

Clinical Uses of Pancreatic Enzymes

Digestive disturbances	Autoimmune disorders
Pancreatic insufficiency	Rheumatoid arthritis
Cystic fibrosis	Lupus erythematosus
Food allergies	Scleroderma
	Multiple sclerosis
Various cancers	Viral infections
Sports injuries and trauma	Herpes zoster (shingles)
	AIDS

Pancreatic Enzymes as Digestive Aids

The most popular clinical use of pancreatic extracts is for impaired digestion caused by either pancreatic insufficiency or a disease like cystic fibrosis. While cystic fibrosis is a rare inherited disorder, pancreatic insufficiency is thought to be relatively common in the elderly.

While starch and fat digestion can be carried out satisfactorily without the help of pancreatic enzymes, the proteases are critical to proper protein digestion. Incomplete digestion of proteins creates a number of problems for the body, including the development of allergies and formation of toxic substances produced during putrefaction—the breakdown of protein material by bacteria.

As well as being necessary for protein digestion, the proteases serve several other important functions. For example, the proteases, along with other digestive secretions, are largely responsible for keeping the small intestine free from parasites (including bacteria, yeast, protozoa, and intestinal worms).[7] A lack of proteases or other digestive secretions greatly increases an individual's risk of having an intestinal infection, including an overgrowth of the yeast *Candida albicans*.

Assessing Pancreatic Function

Physicians use both physical symptoms and laboratory tests to assess pancreatic function. Common symptoms of pancreatic insufficiency include abdominal bloating and discomfort, gas, indigestion and the passing of undigested food in the stool. For laboratory diagnosis, most nutrition-oriented physicians use the comprehensive stool and digestive analysis. This comprehensive analysis will usually reveal the level of pancreatic enzymes being dumped into the intestines from the pancreas by determining the level of excess fat in the stool, excess nitrogen in the stool, and the presence of any other partially or completely undigested food elements. In addition, the complete stool and digestive analysis will also reveal the health of the bacterial flora, which often reflects the degree of pancreatic function.

Pancreatic Enzymes in Food Allergies

In order for a food molecule to produce an allergic response, it must be fairly large. In studies performed in the 1930s and 1940s, pancreatic enzymes were shown to be quite effective in preventing food allergies.[8] It appears that many practitioners are not aware of, or have forgotten about, these early studies. Typically, individuals who do not secrete enough proteases will suffer from multiple food allergies.

Pancreatic Enzymes in Inflammation

The proteases are important in preventing tissue damage during inflammation and the formation of fibrin clots. Proteases cause an increase in the breakdown of fibrin, a process known as fibrinolysis. Fibrin's role in the promotion of inflammation is to form a wall around the area of inflammation, resulting in the blockage of blood and lymph vessels, which leads to swelling. Protease enzymes are often used in the treatment of thrombophlebitis, a disease in which fibrin-caused blood clots develop in veins, which become inflamed, and can dislodge to cause strokes or heart attacks. Pancreatic enzymes and protease enzyme preparations have been shown to be useful in the treatment of many acute and chronic inflammatory conditions, including sports injuries, tendinitis and rheumatoid arthritis.[9,10]

Pancreatic Enzymes in Immune Complex Diseases

The proteases are also essential in preventing the deposit of immune complexes in body tissues. Diseases associated with high levels of circulating immune complexes include rheumatoid arthritis, lupus erythematosus, periarteritis nodosa, scleroderma, ulcerative colitis, Crohn's disease, multiple sclerosis, and AIDS.[11-13] The presence of immune complexes is thought to contribute greatly to the disease process. Experimental and clinical studies have shown that protease enzyme preparations are extremely effective in reducing circulating immune complex levels in several of these diseases. Furthermore, clinical improvements correspond with decreases in immune complex levels.

For example, in the treatment of multiple sclerosis, pancreatic enzyme preparations have been shown to produce good effects in reducing the severity and frequency of symptom flareups.[11] Especially good results were noted in cases of visual disturbance, urinary bladder and intestinal malfunction, and sensory disturbances. However, it should be pointed out that there was little effect on spasticity, dizziness, or tremor noted.

Most of the clinical studies using pancreatic enzyme preparations in inflammatory or autoimmune diseases have utilized enzyme preparations that are relatively weak in potency compared to a number of enzyme preparations available in the United States.

Pancreatic Enzymes in Cancer

One of the most controversial uses of pancreatic enzymes is cancer. Enzyme preparations have been promoted by numerous alternative cancer practitioners for many years, most recently by William Kelley, D.D.S. and Nicholas Gonzalez, M.D. Like most alternative cancer treatments, there is little evidence in the scientific literature to support their use. Several studies have been conducted in Germany, but further research is required.

Dosage

The dosage of pancreatic enzymes is first based on the level of enzyme activity of the particular product. The United States Pharmacopoeia (USP) has set strict definition for level of activity. A 1X pancreatic enzyme (pancreatin) product has in each milligram not less than 25 USP units of amylase activity, not less than 2.0 USP units of lipase activity, and not less than 25 USP units of protease activity. Pancreatin of higher potency is given a whole-number multiple indicating its strength. For example, a full-strength undiluted pancreatic extract that is 10 times stronger than the USP standard would be referred to as 10X USP. Full-strength products are preferred to lower-potency pancreatin products because lower-potency products are often diluted with salt, lactose or galactose to achieve desired strength (e.g., 4X or 1X). The dosage recommendation for a 10X USP pancreatic enzyme product would be 500 to 1,000 mg. three times a day immediately before meals when used as a digestive aid, and 10 to 20 minutes before meals or on an empty stomach when anti-inflammatory effects are desired.

Enzyme products are often enteric-coated; that is, they are coated to prevent digestion in the stomach, so that the enzymes will be liberated in the small intestine. However, numerous studies have shown that non-enteric coated enzyme preparations actually outperform enteric-coated products if they are given prior to a meal (for digestive purposes) or on an empty stomach (for anti-inflammatory effects).

Summary

Pancreatic extracts have proven health benefits. The USP standard allows for easy product comparison. For maximum effects, high-potency preparations (8-10X USP) are preferred.

THYMUS EXTRACTS

The thymus is the major gland of our immune system. It is composed of two soft pinkish-gray lobes lying in a biblike fashion just below the thyroid gland and above the heart. To a very large extent, the health of the thymus determines the health of the immune system. Individuals who get frequent infections or suffer from chronic infections typically have impaired thymus activity. Also, people affected with hay fever, allergies, migraine headaches and rheumatoid arthritis usually have altered thymus function. Before discussing how to support the thymus gland, let's first examine the important roles of the thymus.

Functions of the Thymus

The thymus is responsible for many immune system functions, including the production of T lymphocytes, a type of white blood cell responsible for cell-mediated immunity, that is, immune mechanisms not controlled or mediated by antibodies.

Cell-mediated immunity is extremely important in the resistance to infection by moldlike bacteria, yeast (including *Candida albicans*), fungi, parasites, and viruses (including *Herpes simplex*, Epstein-Barr and viruses that cause hepatitis). Cell-mediated immunity is also critical in protecting against the development of cancer, auto-immune disorders like rheumatoid arthritis, and allergies.

The thymus gland also releases several hormones such as thymosin, thymopoietin and serum thymic factor, which regulate many immune functions. Low levels of these hormones in the blood are associated with depressed immunity and an increased susceptibility to infection. Typically, thymic hormone levels will be very low in the elderly, individuals prone to infection, cancer and AIDS patients, and those exposed to undue stress.

Enhancing Thymus Gland Activity

If there is evidence that the thymus gland is not functioning at a sufficient level, it is necessary to employ measures designed to improve thymus function. Promoting optimal thymus gland activity can involve: (1) prevention of thymic involution or shrinkage by insuring adequate dietary intake of antioxidant nutrients like carotenes, vitamin C, vitamin E, zinc and selenium; (2) use of nutrients (e.g., zinc, vitamins B6 and C) required in the manufacture or action of thymic hormones; and (3) using products containing concentrates of calf thymus tissue.

Therapeutic Uses of Thymus Extracts

There is a substantial amount of clinical data to support the effectiveness of orally administered thymus extracts. Specifically, numerous clinical trials have shown that oral administration of predigested calf thymus extract rich in thymus-derived polypeptides is effective in: (1) preventing recurrent respiratory infections in children; (2) correcting the T-cell defects in human immunodeficiency virus infections (AIDS); (3) treating acute hepatitis B infections; (4) restoring the number of peripheral leukocytes in cancer patients with chemotherapy-induced depression of WBC (white blood cell) counts; and (5) allergies, including asthma, hayfever, and food allergies in children.[1,2]

The effectiveness of the thymus extract in these conditions is reflective of broad-spectrum immune system enhancement presumably mediated by improved thymus gland activity.

Thymus Extracts in Chronic Viral Infections

Recurrent or chronic infections, including the so-called "chronic fatigue syndrome" and "chronic post-viral syndrome," are characterized by a depressed immune system. What makes it difficult to overcome these illnesses is a repetitive cycle: a compromised immune system leads to infection, infection leads to damage to the immune system, further weakening resistance. Thymus extracts may provide the answer to chronic infections by restoring healthy immune function.

The ability of thymus extracts to treat and then reduce the number of recurrent infections was studied in groups of children with a history of recurrent respiratory tract infections. Double-blind studies revealed that orally administered thymus extracts were not only able to effectively eliminate infection, but that treatment over the course of a year significantly reduced the number of respiratory infections and significantly improved numerous immune parameters.[3]

One of the most difficult viral infections for the body to throw off is type B viral hepatitis, but several double-blind studies have shown thymus extracts to be effective in both acute and chronic cases. In these studies, therapeutic effect was noted by accelerated decreases of liver enzymes (transaminases), elimination of the virus and a higher rate of seroconversion to anti-HBe, signifying clinical remission.[4,5]

The most extreme example of a chronic viral infection is AIDS. Although thymus extracts have not shown any real therapeutic benefit in AIDS, studies have shown an ability to improve several immune parameters, including an ability to raise the number of T helper cells, a critical goal in AIDS treatment.[6]

Thymus Extracts in Cancer

The primary application of thymus extracts in cancer has been to counteract the immune-suppressing effects of radiation and chemotherapy. The net effect of thymus extract administration is to prevent the tremendous depression of white blood cell levels and activity as a result of chemotherapy or radiation.[2,7]

Thymus Extracts for Allergies

People with allergies typically have derangements in the immune system. Levels of the allergic IgE antibody and allergic white blood cells known as eosinophils are typically elevated while levels of suppressor T cells are typically depressed. These abnormalities are obviously a result of altered immune function.

The oral administration of thymus extracts has been shown in double-blind clinical studies to improve the symptoms and course of hay fever, allergic rhinitis, asthma, eczema and food allergies.[2,8–10] Presumably this clinical improvement is the result of restoration of proper immune function as levels of IgE and eosinophils have been reduced while the ratio of helper to suppressor T cells has improved.

It is interesting to note that, as shown in several double-blind studies, children receiving thymus extracts during food allergy elimination diets are often able to tolerate foods that had previously been allergenic and symptom-producing.[12, 13]

Thymus Extracts and Autoimmune Disorders

Autoimmune disorders like rheumatoid arthritis are characterized by the body's own antibodies attacking body tissues. Central to this autoimmunity is a high T helper-to-suppressor cell ratio, which results in increased antibody formation. The higher the ratio, the higher the number of antibodies being produced to damage body structures. In one clinical study, rheumatoid arthritis patients with a helper-to-suppressor ratio of 3.3 achieved normal ratios (1.02-2.46) after three months of therapy with a thymus extract.[4]

Although use of a thymus extract may not result in clinical improvement, it appears to be useful in restoring proper immune function in autoimmune diseases including rheumatoid arthritis, lupus and scleroderma.

Dosage

The dosage will vary from one manufacturer to another, as there are no quality control procedures or standards enforced in the glandular industry, it is left up to the individual company to adopt quality control and good manufacturing procedures.

From a practical view, products concentrated and standardized for polypeptide content are preferable to crude preparations. Based on current clinical research, the daily dose should be equivalent to 120 mg. pure polypeptides with molecular sizes less than 10,000 daltons or roughly 500 mg. of the crude polypeptide fraction.

No side effects or adverse effects have been reported with the use of thymus preparations.

Summary

In the attempt to strengthen immunity or in any case of immune system dysfunction, whether it's frequent infection, allergy, or chronic disease, supporting the thymus gland is of prime importance. In addition to supporting the thymus with proper nutrition, the use of preparations containing calf thymus tissue can be extremely helpful in restoring healthy immune system function.

SPLEEN EXTRACTS

The spleen is the largest mass of lymphatic tissue in the body. Weighing about 7 ounces, the spleen is a fist-sized, spongy, dark purple organ that lies in the upper left abdomen behind the lower ribs. The spleen's functions include producing white blood cells, engulfing and destroying bacteria and cellular debris, and destroying worn-out red blood cells and platelets. The spleen also serves as a blood reservoir. During times of demand, such as hemorrhage, the spleen can release its stored blood and prevent shock.

Therapeutic Uses of Spleen Extracts

As early as the 1930s, orally administered bovine spleen extracts were shown to possess some physiological action in increasing white blood cell counts in extreme deficiencies of white blood cells, as well as being of some benefit in patients with malaria and typhoid fever.[1-3]

Like thymus extracts, pharmaceutical grade bovine spleen extracts are currently quite popular in Germany for the treatment of infectious conditions and as an immune-enhancing agent in cancer. Spleen tissue extracts may be of benefit in enhancing general immune function, as many potent immune system enhancing compounds secreted by the spleen are small molecular weight peptides; for example the potent immunostimulants tuftsin and splenopentin are composed of only four and five amino acids, respectively.

Both tuftsin and splenopentin have been shown to exert profound immune-enhancing activity. Tuftsin stimulates specialized white blood cells known as macrophages. Macrophages are actually blood monocytes that have taken up residence in specific tissues like the liver, spleen, and lymph nodes. These large cells engulf and destroy foreign particles, including bacteria, cancer cells and cellular debris. Macrophages are essential in protecting against invasion by microorganisms as well as cancer. Tuftsin also helps mobilize other white blood cells to fight against infection and cancer. A deficiency of tuftsin is associated with signs and symptoms of frequent infections.[4]

Splenopentin, like tuftsin, has also demonstrated significant immune enhancing effects. Its effects are primarily directed towards enhancing the immune system's response to regulating compounds known as "colony-stimulating factors."[5] Colony-stimulating factors like interleukin-3 and granulocyte/macrophage colony-stimulating factors stimulate the production of white blood cells. Splenopentin is probably the factor responsible for the results noted in the 1930s studies mentioned earlier.

Splenopentin has also been shown to enhance natural killer cell activity.[6] Natural killer cells received their name because of their ability to destroy cells that have become cancerous or infected with viruses. They are the body's first line of defense against cancer development.

In addition to tuftsin and splenopentin, hydrolyzed (predigested) spleen extracts concentrated for peptides have demonstrated impressive immune restorative properties in mice.[7] The mice were exposed to radiation, a process which leads to significant destruction of the immune system. Mice treated with the spleen extract recovered within 6 to 8 weeks. In contrast, those treated with a placebo recovered after 10 weeks at the earliest.

The Importance of Spleen Extracts After Splenectomy

A splenectomy is the surgical removal of the spleen. It is usually performed after the spleen has been seriously injured, causing severe hemorrhage. It is necessary to remove the spleen after significant trauma because it is difficult to repair. The spleen is also removed in the medical treatment of certain diseases such as idiopathic thrombocytopenic purpura and to determine the extent of Hodgkin's disease.

The removal of the spleen is associated with an increased risk for infection, leading to susceptibility to pneumococcal pneumonia. About 2.5% of patients having their spleen removed will die from pneumococcal pneumonia within five years after splenectomy. It is often recommended that a child who has undergone a splenectomy receive a pneumococcal vaccine and long-term antibiotic treatment. Use of spleen extracts, especially those rich in tuftsin, may be a natural alternative.

Spleen extracts should probably be viewed as a necessary medicine for people who have had their spleens removed. If your thyroid, adrenals or ovaries were removed, you would be prescribed the corresponding hormone. It only makes sense that if you have your spleen removed, you should supply your body with necessary spleen substances like tuftsin and splenopentin.

The increased risk of infection is attributed primarily to a deficiency of tuftsin.[8,9] Tuftsin is produced only in the spleen; without the spleen there simply is no tuftsin in the circulation. Without tuftsin, the body is missing one of its key stimulators of the immune system. Individuals without spleens need an outside source of tuftsin like spleen extracts.

Another condition where tuftsin levels are quite low is AIDS (acquired immunodeficiency syndrome).[10] A deficiency of tuftsin is linked to the increased risk of bacterial infection in AIDS patients.

Dosage

From a practical view, hydrolyzed (predigested) products concentrated for tuftsin and splenopentin content are preferable to crude preparations. Based on current clinical research, the daily dose should provide 50 mg. tuftsin and splenopentin or roughly 1.5 g. of total spleen peptides. No side effects or adverse effects have been reported with the use of oral spleen preparations.

Summary

The spleen plays a vital role in immune function, largely through the effects of tuftsin and splenopentin. Spleen extracts may be useful in the treatment of low white blood cell counts, bacterial infections, and as an adjunct in cancer therapy. Individuals who have had their spleens removed or who have low tuftsin levels should use spleen extracts.

LIVER EXTRACTS

The liver is truly an intricate, complex and remarkable organ. It is without question the most important organ of metabolism. To a very large extent the health and vitality of an individual is determined by the health and vitality of the liver.

The liver's basic functions are threefold: vascular, secretory, and metabolic. Its vascular functions include being a major blood reservoir and filtering over one quart of blood per minute. The liver effectively removes bacteria, endotoxins, antigen-antibody complexes and various other particles from the circulation.

The liver's secretory functions involve the synthesis and secretion of bile. Each day the liver manufactures about one quart of bile. Bile is necessary for the absorption of fat-soluble substances, including many vitamins. Although the majority of the bile secreted into the intestines is reabsorbed, many toxic substances are effectively eliminated from the body by the bile.

The metabolic functions of the liver are manifold, as the liver is intricately involved in carbohydrate, fat and protein metabolism; the storage of vitamins and minerals; the formation of numerous physiologic factors; and the detoxification or excretion into the bile of various chemical compounds. These include hormones (such as thyroxine, cortisol, estrogen, and aldosterone), histamine, drugs and pesticides.

Liver a Great Source of Iron

Liver is the richest natural source of many vitamins and minerals, including iron. Iron is critical to human life. It plays the central role in the hemoglobin molecule of our red blood cells (RBC) where it functions in transporting both oxygen from the lungs to the body's tissues and carbon dioxide from the tissues to the lungs. Iron also functions in several key enzymes in energy production and metabolism, including DNA synthesis.[1,2]

Iron deficiency is the most common nutrient deficiency in the United States. The groups at highest risk for iron deficiency are infants under two years of age, teenage girls, pregnant women and the elderly. It is estimated that iron deficiency affects 30 to 50 percent of people in these groups. During pregnancy, the number is even higher.[1,2]

Consequences of Iron Deficiency

The negative effects of iron deficiency are due largely to the impaired delivery of oxygen to the tissues and the impaired activity of iron-containing enzymes in various tissues. Iron deficiency can lead to anemia, excessive menstrual loss, learning disabilities, impaired immune function and decreased energy levels and physical performance.[1,2,3]

Anemia is a condition in which the blood is deficient in red blood cells or the hemoglobin (iron-containing) portion of red blood cells. The primary function of the red blood cell is to transport oxygen from the lungs to the tissues of the body in exchange for carbon dioxide. The symptoms of anemia, such as extreme fatigue, reflect a lack of oxygen being delivered to tissues and a buildup of carbon dioxide.

Iron deficiency is the most common cause of anemia, but it must be pointed out that anemia is the last stage of iron deficiency. Iron-dependent enzymes involved in energy production and metabolism are the first to be affected by low iron levels.[2] Serum ferritin is the best laboratory test for determining body iron stores.

Iron and Physical Performance

Several researchers have clearly demonstrated that even slight iron-deficiency anemia leads to a reduction in physical work capacity and productivity.[4-6] Nutrition surveys done in the United States have indicated that iron deficiency is a major cause of impairment of health and work capacity, and thus significant economic loss to the individual and the country. Supplementation with iron has brought about rapid improvements in work capacity in iron-deficient individuals.

Impairment of physical performance by iron deficiency does not necessarily involve anemia. As I just noted, the iron-dependent enzymes involved in energy production and metabolism will be impaired long before anemia occurs.

Understanding Dietary Sources of Iron

There are two forms of dietary iron, "heme" iron and "nonheme" iron. Heme iron is iron bound to hemoglobin and myoglobin. It is the most efficiently absorbed form of iron. The absorption rate of nonheme iron supplements such as ferrous sulfate and ferrous fumarate is 2.9 percent on an empty stomach and 0.9 percent with food. This is much less than the absorption rate of heme iron, as found in liver, which is as high as 35 percent.[1,2] In addition, heme iron is without the side effects associated with nonheme sources of iron such as nausea, flatulence and diarrhea.

Iron Supplements

Iron supplementation is often required, especially during pregnancy and in young menstruating women. Liver extracts provide the best form of iron—heme iron—and are therefore considered by many experts to be the best iron supplements. Liver extracts contain not only iron but also other nutritional and physiological substances which promote healthy red blood cells. Many liver extracts provide the benefit of heme iron without the calories and fat that accompany heme iron from meats.

Liver is a rich source of other nutrients critical in building blood, including vitamin B12 and folic acid. These nutrients also function in maintaining a healthy nervous system and increasing energy levels.

The Endurance-Enhancing Effects of Liver

The mysterious effects of liver have puzzled researchers and scientists for decades. It is thought that liver contains a physiological substance, not yet identified, which promoted stamina, strength and endurance. These attributes have made liver preparations extremely popular among bodybuilders and athletes.

Most clinical studies have utilized hydrolyzed liver extracts produced by exposing the liver material to enzymes which hydrolyze (add water to) the protein bonds. This process basically "liquefies" the liver fraction and is the reason hydrolyzed extracts are often referred to as "liquid liver extracts."

Liver in Liver Disorders

Hydrolyzed liver extracts have long been used to improve liver function. In fact, they have been used in the treatment of many chronic liver diseases since 1896. Numerous scientific investigations into the therapeutic efficacy of liver extracts have demonstrated that these extracts possess an ability to improve fat utilization, promote tissue regeneration and prevent damage to the liver.[7–10] In short, clinical studies have demonstrated that oral administration of liquid liver extracts can be quite effective in improving liver function.

For example, in one double-blind study, 556 patients with chronic hepatitis were given either 70 mg. of liver hydrosylate or a placebo three times daily.[10] At the end of three months of treatment, those receiving the liver extract were shown to have far lower liver enzyme levels than the placebo group. Since the level of liver enzymes in the blood reflects damage to the liver, it can be concluded that the liver extract is effective in chronic hepatitis through its abilities to improve the function of damaged liver cells and to prevent further damage to the liver.

One of the key benefits from the use of liver extracts is that, since the liver is the most important organ of metabolism, when liver function improves, metabolism improves. An improved metabolism creates high energy levels and greater feelings of health and well-being. Liquid liver extracts are a remarkable tonic for well-being.

Dosage

The dosage is entirely dependent upon the concentration, method of preparation, and quality of the liver extract. The highest quality liver extracts are hydrolyzed (liquid) extracts, because they have had the fat-soluble components removed and typically contain more than 20 times the nutritional content of raw liver including 3 to 4 mg of heme iron per gram.

Summary

Liver is rich in many nutrients, including heme iron. The stamina and endurance-enhancing effects of liver are of near-legendary proportions. Hydrolyzed (predigested or "liquid") liver preparations may offer the greatest benefit, especially for those who wish to improve liver function.

ADRENAL EXTRACTS

The adrenal glands lie just above the kidneys and are composed of two distinct parts, the adrenal medulla and the adrenal cortex. The adrenal medulla is functionally related to the sympathetic nervous system, and secretes the hormones epinephrine (adrenalin) and norepinephrine (noradrenaline). These hormones stimulate many body processes related to the flight-or-fight response. They also serve to maintain normal nervous control over many involuntary bodily functions such as heart rate, respiration and digestion.

The adrenal cortex secretes an entirely different group of hormones called corticosteroids. These hormones are all formed from cholesterol. Although all corticosteroids have similar chemical formulas, they differ in function. The three major types of corticosteroids are mineralocorticoids, glucocorticoids, and 17-ketosteroids (sex hormones).

The glucocorticoids, mainly cortisol, corticosterone and cortisone, exert a profound effect upon the metabolism of glucose. These hormones increase serum glucose. In addition, glucocorticoids reduce inflammation and the allergic response.

The mineralocorticoids, of which aldosterone is the most important, have a profound effect on minerals. Specifically, aldosterone increases the retention of sodium and the excretion of potassium by the body.

The 17-ketosteroids are also secreted by the adrenals. The primary sex hormone produced by the adrenal is the androgen (male hormone) dehydroepiandrosterone (DHEA).

Stress and the Adrenals

Some basic control mechanisms operate to counteract the everyday stresses of life. However, if stress is extreme, unusual or long-lasting, these control mechanisms can be quite harmful. Stress triggers a number of biological changes known collectively as the "general adaptation syndrome." The three phases of the general adaptation syndrome are alarm, resistance, and finally exhaustion.[1] These phases are controlled and regulated by the adrenal glands.

Healthy Adrenal Function: A Key to Dealing with Stress

An abnormal adrenal response, either deficient or excessive hormone release, significantly alters an individual's response to stress. Often the adrenals become "exhausted" as a result of constant demands placed upon them. An individual with adrenal exhaustion may feel "stressed out," tired, and have a reduced resistance to allergies and infection.

Atrophy or shrinking of the adrenal cortex is a common side effect of continual stress and cortisone administration. Due to the importance of the adrenal gland, optimum health is dependent on optimum adrenal function. Adrenal extracts, along with supportive nutrients like vitamin C, potassium and pantothenic acid, are often used by individuals who have been exposed to tremendous stress or who have been taking corticosteroid drugs like prednisone for long periods of time.

Forms of Adrenal Extracts

Oral adrenal extracts have been used in medicine since 1931.[2] Adrenal extracts may be made from the whole adrenal or from just the adrenal cortex. Whole adrenal extracts (usually in combination with essential nutrients for the adrenal gland) are most often used in cases of low adrenal function presenting as fatigue, inability to cope with stress and reduced resistance. Because extracts made from the adrenal cortex contain small amounts of corticosteroids, they are typically used as a "natural" cortisone in severe cases of allergy and inflammation (asthma, eczema, psoriasis, rheumatoid arthritis, etc.).

Dosage

The dosage of adrenal extract will depend upon the quality and potency of the product. The best measure of an effective dose for either preparation may be the level of stimulation experienced. If a high-quality preparation is used, as higher dosages are used (for example, twice the label recommendation) a general stimulatory effect will be noticed, including irritability, restlessness and insomnia. I would suggest starting at the label recommendation and slowly increasing your dosage every two days until you notice the stimulatory effect. Once you notice this effect, simply reduce your dosage to a level just below the level that will produce it.

Summary

The adrenal glands control many body functions and play a critical role in resistance to stress. If an individual has experienced a great deal of stress or has taken corticosteroids for a long period of time, the adrenal glands will shrink and not perform properly. Adrenal extracts can be used in people who are "stressed out," or suffer from fatigue, inflammation or allergy. We can support our adrenal glands by learning how to effectively deal with stress through the regular use of relaxation techniques and exercise.

THYROID EXTRACTS

Since the hormones of the thyroid gland regulate metabolism in every cell of the body, a deficiency of thyroid hormones can affect virtually all body functions. The degree of severity of symptoms in the adult ranges from extremely mild deficiency states which are barely detectable (subclinical hypothyroidism) to severe deficiency states which are life-threatening (myxedema).[1-4]

Diagnosing Hypothyroidism

There is much controversy over the diagnosis of hypothyroidism. Before the use of blood measurements, it was common to diagnose hypothyroidism based on basal body temperature (the temperature of the body at rest) and Achilles tendon reflex time (reflexes are slowed in hypothyroidism). With the advent of sophisticated laboratory measurement of thyroid hormones in the blood, these functional tests of thyroid function fell by the wayside. However, it is now known that the blood tests are not sensitive enough to diagnose milder forms of hypothyroidism. As the most common forms of hypothyroidism are mild the majority of people with hypothyroidism are going undiagnosed.[3–5]

Undiagnosed hypothyroidism is a serious concern, as failure to treat such an underlying condition will reduce the effectiveness of nutritional therapies. For example, in most cases zinc, vitamin A and essential fatty acids are effective in relieving dry, scaly skin. However, if a person has hypothyroidism, no improvement will occur. It is critical that thyroid function be evaluated, as hypothyroidism appears to be an underlying factor in a large number of diseases.

The basal body temperature is perhaps the most sensitive functional test of thyroid function.[3,4] A simple method for taking your basal body temperature is detailed below.

Taking Your Basal Body Temperature

Your body temperature reflects your metabolic rate, which is largely determined by hormones secreted by the thyroid gland. The function of the thyroid gland can be determined by simply measuring your basal body temperature. All that is needed is a thermometer.

Procedure

1. Shake down the thermometer to below 95° F and place it by your bed before going to sleep at night.
2. On waking, place the thermometer in your armpit for a full 10 minutes. It is important to make as little movement as possible. Lying and resting with your eyes closed is best. Do not get up until the 10-minute test is completed.
3. After 10 minutes, read and record the temperature and date.
4. Record the temperature for at least three mornings (preferably at the same time of day) and give the information to your physician. Menstruating women must perform the test on the second, third, and fourth days of menstruation. Men and postmenopausal women can perform the test at any time.

Interpretation

Your basal body temperature should be between 97.6° and 98.2°. Low basal body temperatures are quite common and frequently reflect hypothyroidism. Common signs and symptoms of hypothyroidism are low basal body temperature, depression, difficulty in losing weight, dry skin, headaches, lethargy or fatigue, menstrual problems, recurrent infections, constipation and sensitivity to cold.

High basal body temperatures (above 98.6°) are less common, but may be evidence of hyperthyroidism. Common signs and symptoms of hyperthyroidism include bulging eyeballs, fast pulse, hyperactivity, inability to gain weight, insomnia, irritability, menstrual problems and nervousness.

Most estimates on the rate of hypothyroidism are based on the incidence of low levels of thyroid hormone in the blood. As already mentioned, this may mean that a large number of people with mild hypothyroidism go undetected. Nonetheless, even using blood levels of thyroid hormones as the criteria, it is estimated that between 1 percent and 4 percent of the adult population have moderate to severe hypothyroidism, and another 10 percent to 12 percent have mild hypothyroidism.[2, 6-8] The rate of hypothyroidism increases steadily with advancing age.

Some writers of popular books, using medical history, physical examination, and basal body temperatures along with the blood thyroid levels as the diagnostic criteria, estimate the rate of hypothyroidism in the general adult population to be approximately 40 percent.[2,3] It is likely that the true rate of hypothyroidism using these criteria is somewhere near 25 percent of the population.

Manifestations of Adult Hypothyroidism

Since thyroid hormone affects every cell of the body, a deficiency will usually result in a large number of signs and symptoms. A brief review of the common manifestations of hypothyroid on several body systems is given below.

Metabolic: The metabolic manifestations of hypothyroidism reflect a general decrease in the rate of utilization of fat, protein and carbohydrate. Moderate weight gain combined with cold intolerance is a common finding.

Cholesterol and triglyceride levels are increased in even the mildest forms of hypothyroidism.[9] This elevation greatly increases the risk of serious cardiovascular disease. Studies have shown an increased rate of heart disease due to atherosclerosis in individuals with hypothyroidism.[10, 11]

Hypothyroidism also leads to increased capillary permeability and slow lymphatic drainage. Often this will result in swelling of tissue (edema).

Endocrine: A variety of hormonal symptoms can exist in hypothyroidism. Perhaps the most common are a loss of libido (sexual drive) in men and menstrual abnormalities in women.

Women with mild hypothyroidism have prolonged and heavy menstrual bleeding, with a shorter menstrual cycle. Infertility may also be a problem. If the hypothyroid woman does become pregnant, miscarriages, premature deliveries and stillbirths are common. Rarely does a pregnancy terminate in normal labor and delivery in the hypothyroid woman.

Skin, hair and nails: Dry, rough skin covered with fine superficial scales is seen in most hypothyroid individuals, while the hair is course, dry and brittle. Hair loss can be quite severe. The nails become thin and brittle and typically show transverse grooves.

Psychological: The brain appears to be quite sensitive to low levels of thyroid hormone. Depression along with weakness and fatigue are usually the first symptoms of hypothyroidism.[5, 12] Later the hypothyroid individual will have difficulty concentrating and be extremely forgetful.

Muscular and skeletal: Muscle weakness and joint stiffness is a predominant feature of hypothyroidism.[13] Some individuals with hypothyroidism may also experience muscle and joint pain and tenderness.[14]

Cardiovascular: Hypothyroidism is thought to predispose one to atherosclerosis, due to the increase in cholesterol and triglycerides.[9–11] Hypothyroidism can also cause hypertension, reduce the function of the heart and reduce heart rate.

Other manifestations: Shortness of breath, constipation and impaired kidney function are some of the other common features of hypothyroidism.[2]

Dietary Considerations in Hypothyroidism

The manufacture of thyroid hormones within the thyroid gland is dependent on several important nutrients. Deficiency of any of a number of vitamins and minerals, especially iodine, or ingestion of certain foods could result in hypothyroidism.

Iodine and tyrosine: Thyroid hormones are made from iodine and the amino acid tyrosine. A deficiency of iodine results in the development of a goiter—an enlarged thyroid gland. When the level of iodine is low in the diet and blood, it causes the cells of the thyroid gland to become quite large and form a goiter.

Goiter is estimated to affect over 200 million people the world over.[1, 15] In all but 4 percent of these cases the goiter is caused by an iodine deficiency. Iodine deficiency is now quite rare in the United States and other industrialized countries since the addition of iodine to table salt. This practice began in Michigan, where in 1924 the goiter rate was a horrifying 47 percent.

Few people in the United States are now considered iodine deficient, yet the rate of goiter is still relatively high (5 percent to 6 percent) in certain high-risk areas. The goiter in these people is probably a result of the ingestion of certain foods which block iodine utilization. These foods are known as goitrogens and are discussed below.

The recommended dietary allowance (RDA) for iodine in adults is quite small, 150 micrograms.[15] Seafoods, including seaweeds like kelp, clams, lobsters, oysters, and sardines and other saltwater fish, are nature's richest sources of iodine. However, the majority of dietary iodine comes from iodized salt (70 micrograms of iodine per gram of salt). Sea salt has little iodine. The average intake of iodine in the United States is estimated to be over 600 micrograms per day.

Too much iodine can actually inhibit thyroid gland synthesis. For this reason and because the only function of iodine in the body is for thyroid hormone synthesis, it is recommended that dietary levels or supplementation of iodine not exceed 1 milligram (1,000 micrograms) per day for any length of time.

Goitrogens: Some foods contain substances which prevent the utilization of iodine. These foods are termed goitrogens and include turnips, cabbage, mustard, cassava root, soybean, peanuts, pine nuts and millet. Cooking usually inactivates goitrogens.[15]

Vitamins and Minerals: Zinc, vitamin E and vitamin A function together in many body processes, including the manufacture of thyroid hormone.[16] A deficiency of any of these nutrients would result in lower levels of active thyroid hormone being produced. Low zinc levels are common in the elderly, as is hypothyroidism.[17] There may be a correlation.

The B vitamins riboflavin (B2), niacin (B3) and pyridoxine (B6) and vitamin C are also necessary for normal thyroid hormone manufacture.[16]

Correcting Hypothyroidism

The medical treatment of hypothyroidism, in all but its mildest forms, involves the use of desiccated thyroid or synthetic thyroid hormone. Although synthetic hormones have become popular, many physicians (particularly naturopathic physicians) still prefer desiccated natural thyroid, complete with all thyroid hormones. At this time, it appears that thyroid hormone replacement is necessary in the majority of people with hypothyroidism.

The thyroid extracts sold in health food stores are required by the Food and Drug Administration (FDA) to be thyroxine-free. However, it is nearly impossible to remove all the hormone from the gland. In other words, think of health food store thyroid preparations as milder forms of desiccated natural thyroid. If you have mild hypothyroidism, these preparations may provide enough support to help you with your thyroid problem.

Since it is important to support the thyroid gland by insuring adequate intake of key nutrients required in the manufacturing of thyroid hormone and avoiding goitrogens, most health food store thyroid products also contain supportive nutrients like iodine, zinc and tyrosine.

The dosage of commercial thyroid preparations really depends on the potency and level of supportive nutrients. Obviously, these factors will vary from one manufacturer to the next. Therefore, it is best to follow the manufacturer's recommendations as provided on the product's label. Use your basal body temperature to determine effectiveness of the product.

Summary

Hypothyroidism is thought to be a common occurrence in the United States. Determining your basal body temperature is a useful test of thyroid function. In milder forms of hypothyroidism, commercial forms of thyroid available in health food stores may provide enough thyroid hormone and essential nutrients to support healthy thyroid function. In moderate to severe hypothyroidism, thyroid hormone replacement therapy is necessary.

CARTILAGE EXTRACTS, CHONDROITIN SULFATE AND GLUCOSAMINE

Cartilage serves an important role in joint function. Its gel-like nature provides protection to the ends of joints by acting as a shock absorber. Degeneration of the cartilage is the hallmark feature of osteoarthritis (also known as degenerative joint disease). Osteoarthritis is the most common form of arthritis and is seen primarily, but not exclusively, in the elderly. Surveys have indicated that over 40 million Americans have osteoarthritis, including 80 percent of persons over the age of 50.

The cumulative effect of decades of use leads to degenerative changes in joints. This damage is compounded by a decreased ability to restore and manufacture cartilage. The weight-bearing joints, like the knees and hips and joints of the hands are the joints most often affected with osteoarthritis. In affected joints, there is much cartilage destruction followed by hardening and the formation of large bone spurs in the joint margins. Pain, deformity, and limitation of motion in the joint results.

Cartilage Extracts and Glucosamine Sulfate

Cartilage extracts, including chondroitin sulfate, shark cartilage and glucosamine sulfate, are popular supplements in health food stores that may help osteoarthritis by improving cartilage function. These compounds differ in their degree of purity and effectiveness in osteoarthritis.

Shark cartilage contains a mixture of molecules known as glycosaminoglycans (GAGS). One of the key GAGS is chondroitin sulfate. Chondroitin sulfate is composed of repeating units of glucosamine sulfate with attached sugar molecules.

An analogy would be the difference between crude ore (shark cartilage or chondroitin sulfate) and pure gold (glucosamine). While there is gold in crude ore, if you are trying to make jewelry, it is better to use the pure gold. If you are trying to restore cartilage and joint structures, it may be best to use glucosamine sulfate rather than chondroitin sulfate or shark cartilage.

The reason is that glucosamine is much more readily absorbed than shark cartilage and chondroitin sulfate. Shark cartilage and chondroitin sulfate are composed of large molecules that are extremely difficult to absorb. A chondroitin sulfate molecule is 200 times larger than one of glucosamine. While 98 percent of orally administered glucosamine sulfate is absorbed intact, the absorption rate for chondroitin sulfate is estimated to be between zero and 8 percent.[1, 2] Once absorbed, glucosamine is preferentially taken up by cartilage and other joint structures where it simulates the manufacture of mucopolysaccharides like chondroitin sulfate.

From a clinical perspective, glucosamine is extremely effective if given orally. In contrast, the effectiveness of oral chondroitin sulfate and shark cartilage in osteoarthritis is a subject of considerable debate. Most of the studies have utilized injectable forms.[3, 4] If these products are effective when given orally, it may be due to the body breaking down chemical bonds and freeing glucosamine.

Glucosamine Sulfate in Osteoarthritis

There is increasing evidence that the current drug treatment of osteoarthritis is actually doing more harm than good.[4, 5] While aspirin and other drugs like ibuprofen (Motrin, Advil, Nuprin, etc.) are fairly effective in suppressing the symptoms, they actually worsen the condition in the long run because they inhibit cartilage formation and accelerate cartilage destruction.

If current arthritis medications should be avoided, what is an arthritis sufferer to do? It appears that glucosamine sulfate may be nature's best remedy for osteoarthritis.

Glucosamine sulfate is a simple molecule. As its name indicates, it is composed of glucose, an amine (nitrogen and two molecules of hydrogen) and sulfur. In the body, the main action of glucosamine on joints is to stimulate the manufacture of cartilage components, including GAGS. In other words, glucosamine is responsible for stimulating the manufacture of substances necessary for joint repair. This action alone suggests a therapeutic role in osteoarthritis. But there is much more. Glucosamine has also been shown to exert a protective effect against joint destruction and, when given orally as glucosamine sulfate, it is selectively taken up by joint tissues to exert a powerful therapeutic effect in osteoarthritis.

Numerous double-blind studies have shown glucosamine sulfate to yield results as good as or even better than nonsteroidal anti-inflammatory drugs (NSAIDs) in relieving the pain and inflammation of osteoarthritis.[6–9] What is amazing about these results it the fact that glucosamine sulfate exhibits no pain-relieving effects. While NSAIDs offer purely symptomatic relief and may actually promote the disease process, glucosamine sulfate addresses the cause of osteoarthritis. By getting at the root of the problem, glucosamine sulfate not only relieves the symptoms, including pain, it also helps the body repair damaged joints. This effect is outstanding, especially in view of the safety and the lack of side effects associated with oral glucosamine sulfate. In contrast, the side effects and risks associated with drugs currently used in the treatment of osteoarthritis are significant. The therapeutic margin, a measure of safety, is 10 to 30 times more favorable for glucosamine sulfate than for commonly used NSAIDs and no adverse side effects have been reported.[7]

The beneficial results with glucosamine are more obvious the longer it is used. Because glucosamine sulfate is not a pain-relieving substance per se, it takes a while longer than NSAIDs to produce

Comparison of Pain Relief with Glucosamine and Ibuprofen

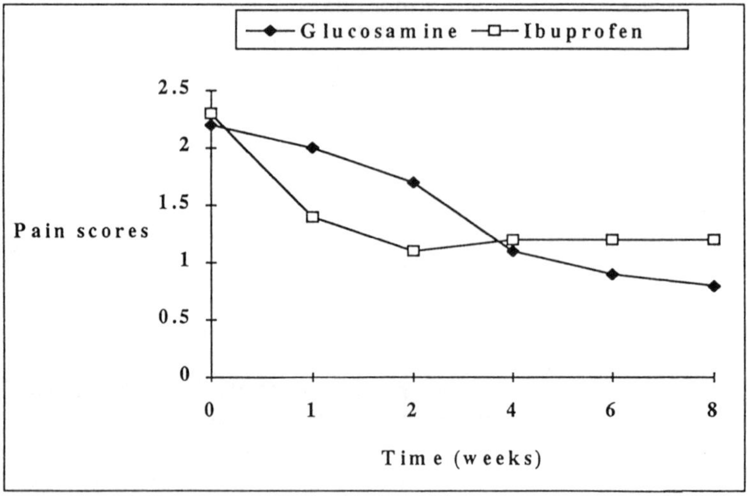

results, but once it starts working, it will produce much better results. For example, in one study that compared glucosamine sulfate to ibuprofen, the active ingredient in Motrin, Advil, and Nuprin, although pain scores decreased faster in the first 2 weeks in the ibuprofen group, by week 4 the group receiving the glucosamine sulfate was doing significantly better than the ibuprofen group.[6]

Glucosamine sulfate products are available at health food stores or through nutritionally-oriented physicians. The standard dose for glucosamine sulfate is 500 mg three times per day. As mentioned earlier, glucosamine sulfate is extremely well tolerated. In addition, there are no contraindications or adverse interactions with drugs. Glucosamine sulfate may cause some gastrointestinal upset (nausea, heartburn, etc.) in rare instances. If this occurs, try taking it with meals.

Shark Cartilage: An Angiogenesis Inhibitor

Shark cartilage, as well as other animal cartilage, contains a substance that strongly inhibits the growth of new blood vessels (angiogenesis) toward solid tumors, thereby restricting tumor growth.[10, 11] Since the aggressiveness of tumor growth is directly related to the number of new blood vessels being extended from the tumors, inhibiting new blood vessel formation is an exciting potential method of treating cancer. It is something like tying to tourniquet around your arm. Eventually you will cut off the blood supply and the arm will start to decompose. Cutting off the blood supply to the tumor might possibly cause it to shrink and disappear.

Most of the studies on cartilage and cancer have been performed in test tubes and on animals, often using injectable forms. In one study of 31 humans with a variety of malignancies, including cancers of the ovary, colon, prostate, cervix and thyroid, oral and injectable bovine tracheal cartilage produced an overall response rate of 90 percent. Therapy was usually started by injection and then an oral dose of 3 grams every eight hours (9 grams per day) was prescribed.[12]

When the popular television program "60 Minutes" ran a segment on shark cartilage, health food stores were besieged by consumers wanting its potential benefits. While shark cartilage may prove to be an effective antitumor therapy, it is unlikely that the thousands of eager consumers using it are taking an effective amount. Studies in Mexico and Cuba demonstrating success of shark cartilage on end-stage cancer patients used a dosage of 1.5 grams per kilogram (2.2 pounds) body weight or roughly 100-125 grams daily.[13] Because of the high dosage, many patients were administered the shark cartilage in the form of a retention enema. More studies are currently under way.

The critical fact to point out regarding the preliminary research on the antitumor effect of cartilage preparations, especially shark cartilage, are that they appear to be most useful with solid tumors, if they are given at high enough dosages. For example, most commercial products provide 750 mg. of shark cartilage per tablet or capsule. This means that to approximate the dosage used in the studies in end-stage cancer patients would require two hundred tablets or capsules per day at a cost of about $50 per day or $1,500 per month.

Until there are more studies, my recommendation to cancer patients is to explore other natural therapies rather than use cartilage products.

Summary

Degeneration of the cartilage leads to osteoarthritis. Although cartilage extracts, including shark cartilage, chondroitin sulfate and green-lipped mussel, may be helpful, in the treatment of osteoarthritis the best choice appears to be glucosamine sulfate.

Animal cartilage is known to inhibit the formation of new blood vessels supplying solid tumors. Although preliminary research with cartilage products is exciting, there remain many questions regarding effectiveness and dosage.

AORTA EXTRACTS

The aorta is the main artery of the body. It arises directly from the heart to supply oxygenated blood to all other arteries of the body. Bovine aorta extracts are chiefly composed of mesoglycan—the name given to describe the mucopolysaccharide or glycosaminoglycan (GAG) complex contained within the arteries of humans and other mammals. Simply stated, glycosaminoglycans are the ground substance components responsible for holding cells of the body together and providing structural support. Mesoglycan functions as structural support to blood vessels. It also acts to protect and promote normal artery and vein function. This is of extreme importance, as many Americans would benefit from improved health of their arteries and veins.

Consistent with the theory behind modern glandular therapy—like cures like—the use of extracts of arteries, including the aorta, may be beneficial. There is considerable evidence of the clinical effectiveness of bovine aorta extracts (mesoglycan) in improving arterial function and blood flow. There is also evidence indicating that aorta extracts address many of the underlying features that contribute to the development of atherosclerosis—the process of hardening of the artery walls caused by a buildup of plaque containing cholesterol, fatty material and cellular debris. If blood flow through an artery is restricted or blocked by atherosclerosis, severe damage or death to the tissues the vessel supplies will result. If the arteries supplying the heart become blocked, the result is what is known as a heart attack. Atherosclerosis and its complications are the major causes of death in the United States and have reached epidemic proportions throughout all of the Western world.

Mesoglycan in Atherosclerosis

Mesoglycan has many important effects which interfere with the progression of atherosclerosis, including preventing damage to the surface of the artery, formation of damaging blood clots, and formation of fat and cholesterol deposits, as well as lowering total cholesterol levels while raising HDL-cholesterol.[1–6]

Studies in animals have indicated that mesoglycan is effectively absorbed orally and is incorporated into vessels where it dramatically improves the integrity and function of arteries.[7]

As for human use, a number of clinical studies have demonstrated that supplementing the diet with mesoglycan has remarkable effect in improving the structure, function, and integrity of arteries as well as improving blood flow.[8–10.] Mesoglycan has been shown to be effective in improving both cerebral (brain) and peripheral (hands and feet) vascular insufficiency. Significant improvements in both symptoms and blood flow have been noted.

Symptoms of cerebral vascular insufficiency can include: short-term memory loss, vertigo, headache, ringing in the ears and depression. These are often referred to as "symptoms of aging" and represent almost entirely a reduced supply of blood and oxygen to the brain due to atherosclerosis. Symptoms of peripheral vascular disease can include: coldness of hands or feet, pain, muscle cramps and (in males) impotence.

The Importance of Mesoglycan in Maintaining the Health of Veins

Veins are fairly frail structures. Defects in the wall of a vein lead to dilation of the vein and damage to the valves. When the valves become damaged, the increased pressure results in the bulging veins known as varicose veins.

Varicose veins affect nearly 50 percent of middle-aged adults. The veins just under the skin of the legs are the veins most commonly affected, due to the tremendous strain that standing puts on these veins. When an individual stands for long periods of time, the pressure exerted against the vein can increase up to ten times. Hence, individuals with occupations that require long periods of standing are at greatest risk for developing varicose veins.

Women are affected about four times as frequently as men; obese individuals have a much greater risk; and the risk increases with age due to loss of tissue tone, loss of muscle mass and weakening of the walls of the veins. Pregnancy may also lead to the development of varicose veins, as pregnancy increases venous pressure in the legs.

Several theories exist to explain the cause of varicose veins: genetic weakness of the veins or venous valves; excessive venous pressure due to a low-fiber induced increase in straining during defecation; long periods of standing, and/or heavy lifting; damage to the veins or venous valves secondary to thrombophlebitis; and weakness of the vascular walls due to either abnormalities in the proteoglycans of the connective tissue or excessive release of cellular enzymes which break down the ground substance, resulting in increased capillary permeability and loss of integrity of the venous structure.

Glycosaminoglycans (GAGS), including mesoglycan, provide the skeletal framework of the vein and are therefore essential in maintaining the structure and integrity of veins. Without proper structural support, veins will lose their shape and become bulging and unsightful.

Aorta extracts (mesoglycan) have demonstrated impressive clinical results in improving the function and structure of veins.[11–16] Individuals with poor venous function of the legs typically experience such symptoms as a sense of heaviness in the legs, tingling sensations, fluid retention, itching, and painful cramps. These symptoms are improved with mesoglycan, due to its ability to improve the structure and function of the vein thereby allowing improved blood flow.

Dosage

The dosage of aorta extract should provide a daily dosage of at least 100 mg of mesoglycan or total aortic glycosaminoglycans. Chondroitin sulfate can be used as an alternative to aorta extracts at a daily dose of 3 grams (1 gram with meals, three times daily).[17] Similar, but not nearly as impressive, results to aorta extracts in the treatment of atherosclerosis have been noted with chondroitin sulfate.

Summary

Bovine aorta extracts rich in glycosaminoglycans (mesoglycan) may be useful in enhancing the function, structure, and integrity of arteries and veins. Mesoglycan has many important effects which interfere with the progression of atherosclerosis including preventing damage to the surface of the artery, formation of damaging blood clots, migration of smooth muscle cells into the intima, and formation of fat and cholesterol deposits, as well as lowering total cholesterol levels while raising HDL-cholesterol.

FINAL COMMENTS

As modern medicine evolves, so will glandular therapy. Currently, there is a tremendous resurgence in the use of glandular extracts in many parts of the world. The application of modern extraction and processing techniques are improving the quality of glandular products available. As with other natural food supplements and herbal products, a higher-quality product translates into high-quality results.

Although there is much research that still needs to be performed on glandular therapy, the basic premise that when an organ/gland is diseased or not functioning properly, supplementing the diet with the corresponding gland of animal origin certainly has scientific support. Like so many time-tested therapies, it appears that glandular therapy's historical use will be confirmed by scientific investigation.

References

Introduction

1. Gardener MLG: Gastrointestinal absorption of intact proteins. Ann Rev Nutr 8:329-50, 1988.
2. Gardner MLG: Intestinal assimilation of intact peptides and proteins from the diet—a neglected field? Biol Rev 59:289-331, 1984.
3. Udall JN and Walker WA: The physiologic and pathologic basis for the transport of macromolecules across the intestinal tract. J. Pediatr. Gastroenterol Nutr 1:295-301, 1982.
5. Hemmings WA and Williams EW: Transport of large breakdown products of dietary protein through the gut wall. Gut 27: 715-23, 1986.
6. Ambrus JL, et al.: Absorption of exogenous and endogenous proteolytic enzymes. Clin Pharmacol Therap 8:362-8, 1967.
7. Kabacoff BB, et al.: Absorption of chymotrypsin from the intestinal tract. Nature 199:815-7, 1963.
8. Ormiston BJ: Clinical effects of TRH and TSH after i.v. and oral administration in normal volunteers and patients with thyroid disease. In: Thytropin Releasing Hormone (Frontiers of Hormone Research, Vol. I), R Hal et al. (eds), Karger, Basel, Switzerland, 1972. pp 45-52.
9. Amoss M, et al.: Release of gonadotrophins by oral administration of synthetic LRF or tripeptide fragment of LRF. J Clin Endocrinol Metab. 35:175-7, 1977.
10. Seifert, J, et al.: Mucosal permeation of macromolecules and particles. Angiology 17:505-13, 1966.
11. Laskowski M, et al.: Effect of trypsin inhibitor on passage of insulin across the intestinal barrier. Science, 127:1115-6 1958.

Pancreatic Extracts

1. Ambrus JL, et al.: Absorption of exogenous and endogenous proteolytic enzymes. Clin Pharmacol Therap 8:362-8, 1967.
2. Kabacoff BB, et al.: Absorption of chymotrypsin from the intestinal tract. Nature 199:815-7, 1963.
3. Martin GJ, et al.: Further in vivo observations with radioactive trypsin. Am J Pharm 129:386-92, 1964.
4. Avakian S: Further studies on the absorption of chymotrypsin. Clin Pharmacol Therap 5:712-5, 1964.
5. Miller JM, et al.: An investigation of trypsin I in patients. Exper Med Surg 18:352-70, 1960.
6. Liebow C and Rothman SS: Enteropancreatic circulation of digestive enzymes. Science 189:472-4, 1975.
7. Rubinstein E, et al.: Antibacterial activity of the pancreatic fluid. Gastroenterol 88:927-32, 1985.
8. Oelgoetz AW, et al.: The treatment of food allergy and indigestion of pancreatic origin with pancreatic enzymes. Am J Dig Dis Nutr 2:422-6, 1935.
9. Innerfield I: Enzymes in Clinical Medicine. McGraw Hill, New York, 1960.
10. Horger I: Enzyme therapy in multiple

rheumatic diseases. Therapiewoche 33:3948-57, 1983.
11. Ransberger K: Enzyme treatment of immune complex diseases. Arthritis Rheuma 8:16-9, 1986.
12. Steffen C, et al.: Enzyme therapy in comparison with immune complex determinations in chronic polyarteritis. Rheumatologie 44:51-6, 1985.
13. Ransberger K and van Schaik W: Enzyme therapy in multiple sclerosis. Der Kassenarzt 41:42-5, 1986.

Thymus Extracts

1. Cazzola P, Mazzanti P and Bossi G: In vivo modulating effect of a calf thymus acid lysate on human T lymphocyte subsets and CD4+/CD8+ ratio in the course of different diseases. Curr Ther Res 42:1011-7, 1987.
2. Kouttab NM, Prada M and Cazzola P: Thymomodulin: Biological properties and clinical applications. Medical Oncology and Tumor Pharmacotherapy 6:5-9, 1989.
3. Fiocchi A, et al.: A double-blind clinical trial for the evaluation of the therapeutic effectiveness of a calf thymus derivative (Thymomodulin) in children with recurrent respiratory infections. Thymus 8:831-9, 1986.
4. Galli M, et al.: Attempt to treat acute type B hepatitis with an orally administered thymic extract (Thymomodulin): Preliminary results. Drugs Exptl Clin Res 11:665-9, 1985
5. Bortolotti F, Cadrobbi P, Crivellaro C, et al.: Effect of an orally administered thymic derivative, Thymodulin, in chronic type B hepatitis in children. Curr Ther Res 43:67-72, 1988.
6. Valesini G, Barnaba V, Benvenuto R, et al.: A calf thymus lysate improves clinical symptoms and T-cell defects in the early stages of HIV infection: Second report. Eur. J Cancer Clin Oncol 23:1915-9, 1987.
7. Kang SD, Lee BH, Yang JH and Lee CY: The effects of calf-thymus extract on recovery of bone marrow function in anticancer chemotherapy. New Med J (Korea) 28:11-5, 1985.
8. Marzari R, Mazzanti P, Cazzola P and Pirodda E: Perennial allergic rhinitis: prevention of the acute episodes with Thymomodulin. Min Med 78:1675-81, 1987.
9. Genova R and Guerra A: Thymomodulin in management of food allergy in children. Int J Tiss React 8:239-42, 1986.
10. Cavagni G, Piscopo E, Rigoli E, et al.: Food allergy in children: an attempt to improve the effects of the elimination diet with an immunomodulating agent (thymomodulin). A double-blind clinical trial. Immunopharmacol Immunotoxicol 11:131-42, 1989.

Spleen Extracts

1. Minter MM: Agranulocytic angina: Treatment of a case with fetal calf spleen. Texas State J Med 2:338-43, 1933.
2. Gray GA: The treatment of agranulocytic angina with fetal calf spleen. Texas State J Med 29:366-9, 1933.
3. Greer AE: Use of fetal spleen in agranulocytosis: Preliminary report. Texas State J Med 28:338-43, 1932.
4. Fridkin M and Najjar VA: Tuftsin: its chemistry, biology, and clinical potential. Crit Rev Biochem Mol Biol 24:1-40, 1989.
5. Diezel W, et al.: The effect of splenopentin (DA SP-5) on in vitro myelopoiesis and on AZT-induced bone marrow toxicity. Int J Immunopharmac 15:269-73, 1993.
6. Rastogi A, et al.: Augmentation of human natural killer cells by splenopentin analogs. FEBS Lett 317:93-5, 1993.
7. Volk HD, et al.: Immunorestitution by a bovine spleen hydrosylate and ultrafiltrate. Arzniem Forsch 41:1281-5, 1991.
8. He SW: Effect of splenectomy on phagocytic function of leukocytes. Chung Hua Wai Ko Tsa Chih 27:354-6,381, 1989.
9. Spirer Z, et al.:Decreased tuftsin concentrations in patients who have un-

dergone splenectomy. Br Med J 2:1574-6, 1977.
10. Corazza GR, et al.: Tuftsin deficiency in AIDS. Lancet i:12-3, 1991.

Liver Extracts

1. Krause MV and Mahan KL: Food, Nutrition and Diet Therapy, 7th ed. WB Saunders, Philadelphia, PA, 1984. pp128-31, 157-64, 585-99.
2. Fairbanks VF and Beutler E: Iron. In: Modern Nutrition in Health and Disease, 7th Ed. Shils ME and Young VR (eds). Lea and Febiger, Philadelphia, PA, 1988 pp193-226.
3. Cook JD and Lynch SR: The liabilities of iron deficiency. Blood 68:803-9, 1986.
4. Viteri FE and Torun B: Anaemia and physical work capacity. Clin Haematol 3:609-26, 1974.
5. Basta SS, Soekirman MS, Karyadi D and Scrimshaw NS: Iron deficiency anemia and the productivity of adult males in Indonesia. Am J Clin Nutr 32:9'6-25, 1979.
6. Gardner GW, Edgerton VR, Senewiratne B, et al.: Physical work capacity and metabolic stress in subjects with iron deficiency anemia. Am J Clin Nutr 30:910-7, 1977.
7. Nagai K: A study of the excretory mechanism of the liver—effect of liver hydrolysate on BSP excretion. Jap J Gastroenterol 67:633-8, 1970.
8. Ohbayashi A, Akioka T and Tasaki H: A study of effects of liver hydrolysate on hepatic circulation. J Therapy 54:1582-5, 1972.
9. Sanbe K, et al.: Treatment of liver disease—with particular reference to liver hydrolysates. Jap J Clin Exp Med 50:2665-76, 1973.
10. Fujisawa K, et al.: Therapeutic effects of liver hydrolysate preparation on chronic hepatitis—A double blind, controlled study. Asian Med J 26:497-526, 1984.

Adrenal Extracts

1. Selye H: Stress in Health and Disease. Buttersworth, London, UK, 1976
2. Britton SW and Silvette H: Further experiments on cortico-adrenal extract: Its efficacy by mouth. Science 74:440-1, 1931.

Thyroid Extracts

1. Petersdorf, R., et al (eds): Harrison's Principles of Internal Medicine. McGraw-Hill, New York, NY, 1983, pp614-23.
2. Mazzaferri EL: Adult hypothyroidism. Postgraduate Medicine 79:64-72, 1986.
3. Barnes BO and Galton L: Hypothyroidism: The Unsuspected Illness. Thomas Crowell, New York, NY, 1976
4. Langer SE and Scheer JF: Solved: The Riddle of Illness. Keats, New Canaan, CT, 1984.
5. Gold M, Pottash A, and Extein I: Hypothyroidism and depression, evidence from complete thyroid function evaluation. JAMA 245:1919-22, 1981.
6. Drinka PJ and Nolten WE: Review: Subclinical hypothyroidism in the elderly: To treat or not to treat? Am J Med Sci 295:125-8, 1988.
7. Banovac K, Zakarija M and McKenzie JM: Experience with routine thyroid function testing: abnormal results in "normal" populations. J Florida Med Assoc 72:835-9, 1985.
8. Rosenthal MJ, Hunt WC, Garry PJ and Goodwin JS: Thyroid failure in the elderly: Microsomal antibodies as discriminate for therapy. JAMA 258:209-13, 1987.
9. Althaus U, Staub JJ, Ryff-De Leche A, et al.: LDL/HDL-changes in subclinical hypothyroidism: possible risk factors for coronary heart disease. Clinical Endocrinology 28:157-63, 1988.
10. Dean JW and Fowler PBS: Exaggerated responsiveness to thyrotrophin releasing hormone: a risk factor in women with artery disease. Brit Med J 290:1555-61, 1985.
11. Turnbridge WMG, Evered DC and Hall R: Lipid profiles and cardiovas-

cular disease in the Wickham area with particular reference to thyroid failure. Clin Endocrinol 7:495-508, 1977.
12. Joffe R, Roy-Byrne P, and Udhe T: Thyroid function and affective illness: a reappraisal. Biol Psychiatry 19:1685-91, 1984.
13. Krupsky M, et al.: Musculoskeletal symptoms as a presenting sign of long-standing hypothyroidism. Isr J Med Sci 23:1110-3, 1987.
14. Hochberg MC, et al.: Hypothyroidism presenting as a polymyositis/like syndrome. Arthr Rheum 19:1363-6, 1976
15. Krause MV and Mahan LK: Food, Nutrition, and Diet Therapy, 7th edition. WB Saunders, Philadelphia, PA, 1984. pp170-4.
16. Jennings IW: Vitamins in Endocrine Metabolism. C.C. Thomas, Springfield, IL, 1970.
17. Prasad A: Clinical, biochemical and nutritional spectrum of zinc deficiency in human subjects: An update. Ntr Rev 41:197-208, 1983.

Cartilage Extracts, Chondroitin Sulfate and Glucosamine

1. Setnikar I, Giachetti C and Zanolo G: Pharmacokinetics of glucosamine in the dog and in man. Arzneim Forsch 36:729-35, 1986.
2. Morrison LM: Therapeutic applications of chondroitin-4-sulfate, appraisal of biologic properties. Folia Angiol 25:225-32, 1977.
3. Prudden JF and Balassa LL. The biological activity of bovine cartilage preparations. Semin Arthritis Rheum 3:287-21, 1974.
4. Brooks PM, Potter SR and Buchanan WW: NSAID and osteoarthritis—help or hindrance. J Rheumatol 9:3-5, 1982.
5. Newman NM and Ling RSM Acetabular bone destruction related to nonsteroidal anti-inflammatory drugs. Lancet: ii; 11-13, 1985.
6. Vaz AL: Double-blind clinical evaluation of the relative efficacy of ibuprofen and glucosamine sulfate in the management of osteoarthrosis of the knee in out-patients. Curr Med Res Opin 8:145-9, 1982.
7. Crolle G and D'este E: Glucosamine sulfate for the management of arthrosis: a controlled clinical investigation. Curr Med Res Opin 7:104-14, 1980.
8. Tapadinhas MJ, Rivera IC and Bignamini AA: Oral glucosamine sulfate in the management of arthrosis: report on a multi-centre open investigation in Portugal. Pharmatherapeutica 3:157-68, 1982.
9. D'Ambrosia ED, Casa B, Bompani R, Scali G and Scali M: Glucosamine sulphate: a controlled clinical investigation in arthrosis. Pharmatherapeutica 2:504-8, 1982.
10. Lee A and Langer R: Shark cartilage contains inhibitors of tumor angiogenesis. Science 221:1185-7, 1983.
11. Folkman J, et al.: Induction of angiogenesis during the transition from hyperplasia to neoplasia. Nature 339:58-61, 1989.
12. Prudden JF: The treatment of human cancer with agents prepared from bovine cartilage. J Biol Response Mod 4:551-84, 1985.
13. Quillin P: Beating Cancer with Nutrition. Nutrition Times Press. Tulsa, OK, 1994, pp114-5.

Aorta Extracts

1. Stevens RL, et al.: The glycosaminoglycans of the human artery and their changes in atherosclerosis. J Clin Invest 58:470-81, 1976.
2. Tammi M, et al.: Connective tissue components in normal and atherosclerotic human coronary arteries. Atherosclerosis 29:191-4, 1978.
3. Day CE, Powell JR and Levy RS: Sulfated polysaccharide inhibition of aortic uptake of low density lipoproteins. Artery 1:126-37, 1975.
4. Pernigotti LM, et al.: Effect of mesoglycan on clotting, fibrinolysis and platelet aggregation in normal subjects and hyperaggregating arteriosclerosis. In: Widhalm K and Sinzinger J, eds. Cur-

rent Aspects of Atherosclerosis. Lipids, lipoproteins, platelets, prostaglandins, and experimental findings. Verlag Wilhelm, Madrich, 1983, pp164-75.
5. Postiglione A, et al.: Effect of oral mesoglycan-sulphate on plasma lipoprotein concentration and on lipoprotein concentration in primary hyperlipidemia. Pharmacol Res Commun 16:1-8, 1984.
6. Saba P, et al.: Hypolipidemic effect of mesoglycan in hyperlipidemic patients. Current Therapeutic Research 40:761-8, 1986.
7. Zanolo G, et al.: Pharmacokinetic aspect of tritium-labeled glycosaminoglycans (mesoglycan) their absorption in rat and monkey and tissue distribution in rat. Boll Chim Farm 123:223-35, 1984
8. Sangrigoli V, et al.: Our experience in therapy with mesoglycan in the treatment of peripheral chronic arterial insufficiency. Proceedings of the IX National Congress of the Italian Society of Vascular Pathology. Copanello, Italy, June 6-9, 1987
9. Mansi D, et al.: Open trial of mesoglycan in the treatment of cerebrovascular ischemic disease. Acta Neurologica 10:108-12, 1988
10. Catalini R, et al.: Echographic monitoring of atheromatous injuries after chronic administration of mesoglycan. Proceedings of the IX National Congress of the Italian Society of Vascular Pathology. Copanello, Italy, June 6-9, 1987
11. Scolari EA: Chronic arterial diseases of the lower limbs: Medical or surgical treatment? Comparative study. Proceedings of the IX National Congress of the Italian Society of Vascular Pathology. Copanello, Italy, June 6-9, 1987
12. De Donato G and Sangiuolo P: Instrumental evaluation of the mesoglycan effects in phlebopathic patients. Prospective randomized double-blind study. Minerva Medica 77:1927-31, 1986
13. Oddone G, Fiscella GF and De Franceschi T: Assessment of the effects of oral mesoglycan sulphate in patients with chronic venous pathology of the lower extremities. Gazzetta Medica Italian 146:111-4, 1987
14. Scondotto G, et al.: Pharmacological effects on venous function evaluated by strain-gauge computerized plethysmography. Proceedings of the IX National Congress of the Italian Society of Vascular Pathology. Copanello, Italy, June 6-9, 1987
15. Castriotta G and Ambroso GA: Efficacy of mesoglycan in the treatment of chronic venous insufficiency of the lower limbs and its complications. Proceedings of the 4th Italian Congress of Phlebology. Naples, Italy November 30-December 3, 1987
16. Gattuso R, et al.: Mesoglycan in the treatment of recurrent phlebitis. Proceedings of the 4th Italian Congress in Phlebology. Naples, Italy November 30-December 3, 1987.
17. Nakazawa K and Murata K: The therapeutic effect of chondroitin polysulphate in elderly atherosclerotic patients. J Int Med Res 6:217-25, 1978.